# Science
# Technology
# Engineering
# Mathematics
# Coloring & Activity
# Book

By Dr. Thomai Dion

ISBN-13: 978-1543271881

# As you color the flower, can you also name its different parts?

Anther

Filament

Petal

Stigma

Pollen

Stem

Style

Ovary

Ovule

Sepal

Receptacle

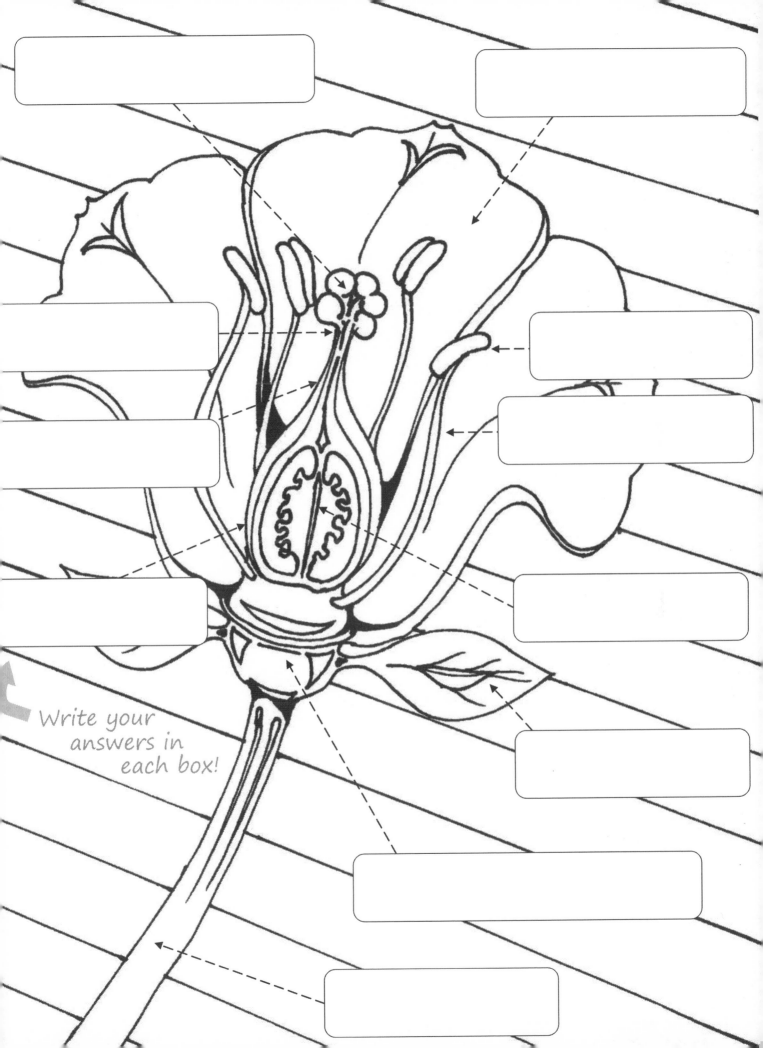

Write your
answers in
each box!

# Binary code

is a numerical system
used by computers that
includes only the numbers
"0" and "1".

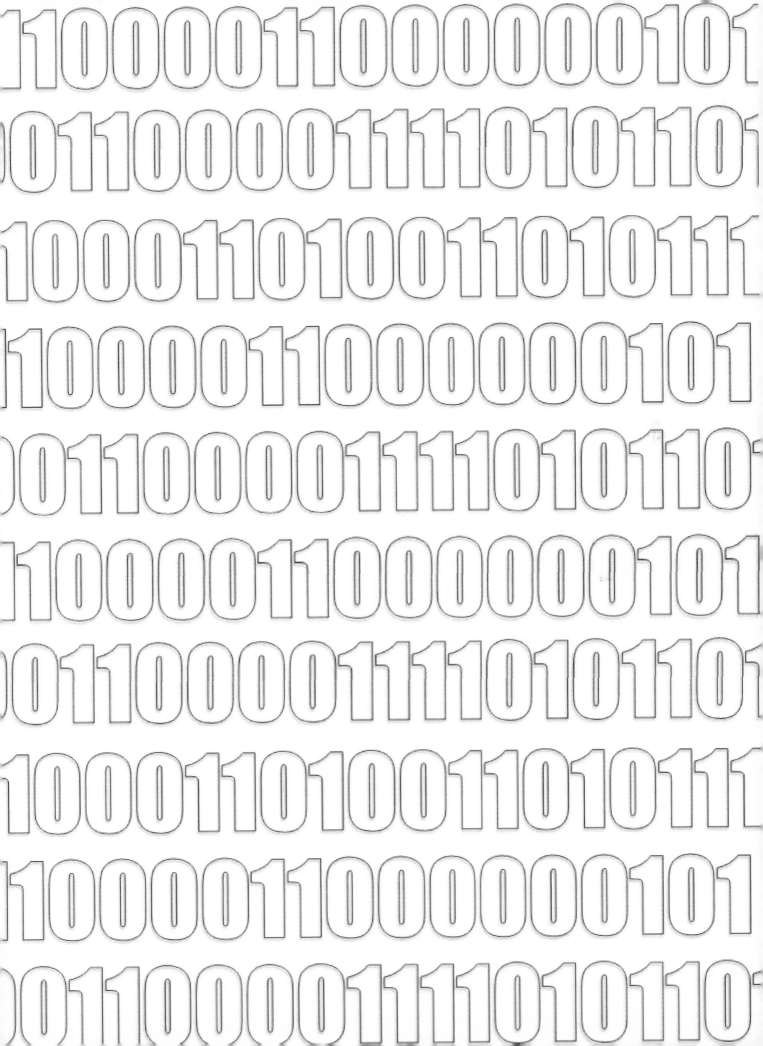

# Neurons

are specialized cells that transmit signals from our brains, allowing our bodies to move.

There are 4 major mathematical operations found in basic arithmetic. These operations are addition, subtraction, multiplication and division.

# The human heart

is an organ that is working all the time. It beats roughly 100,000 times a day to pump blood throughout our bodies and keep us alive.

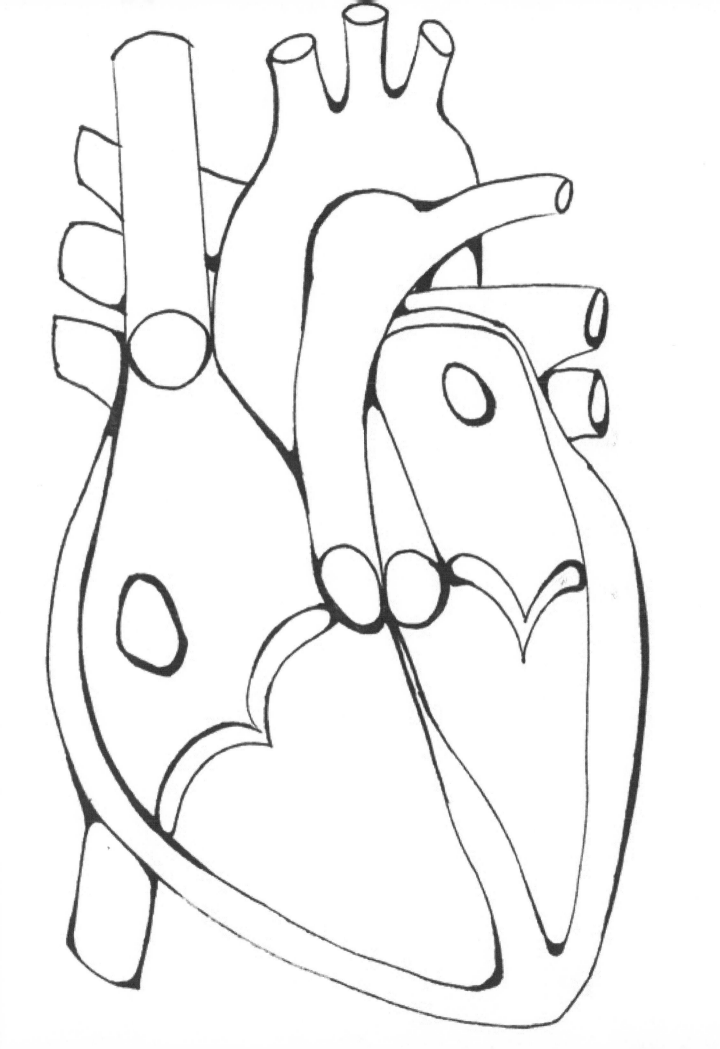

There are many different types of triangles found in geometry. Can you identify these types as you color the next page?

Right triangle
Equilateral triangle
Isosceles triangle
Obtuse triangle

# What does the acronym "S.T.E.M." stand for?

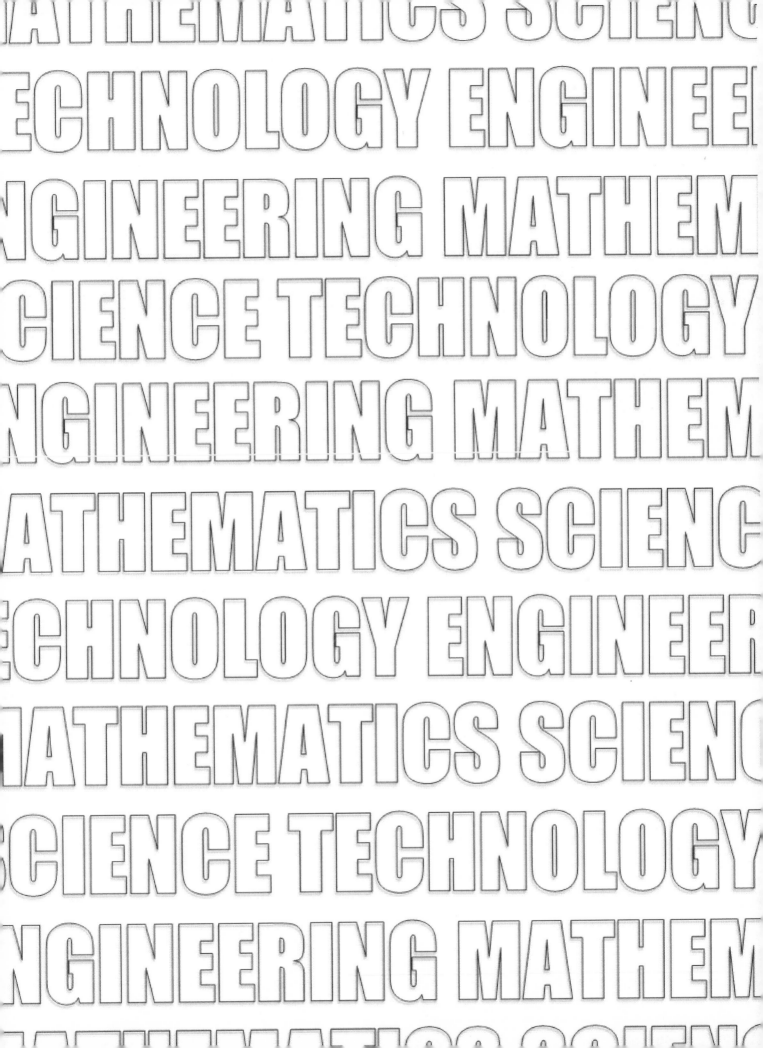

As you color the ant, can you also name the different parts of its body?

Head

Thorax

Abdomen

Antenna

Eye

Leg

Mandible

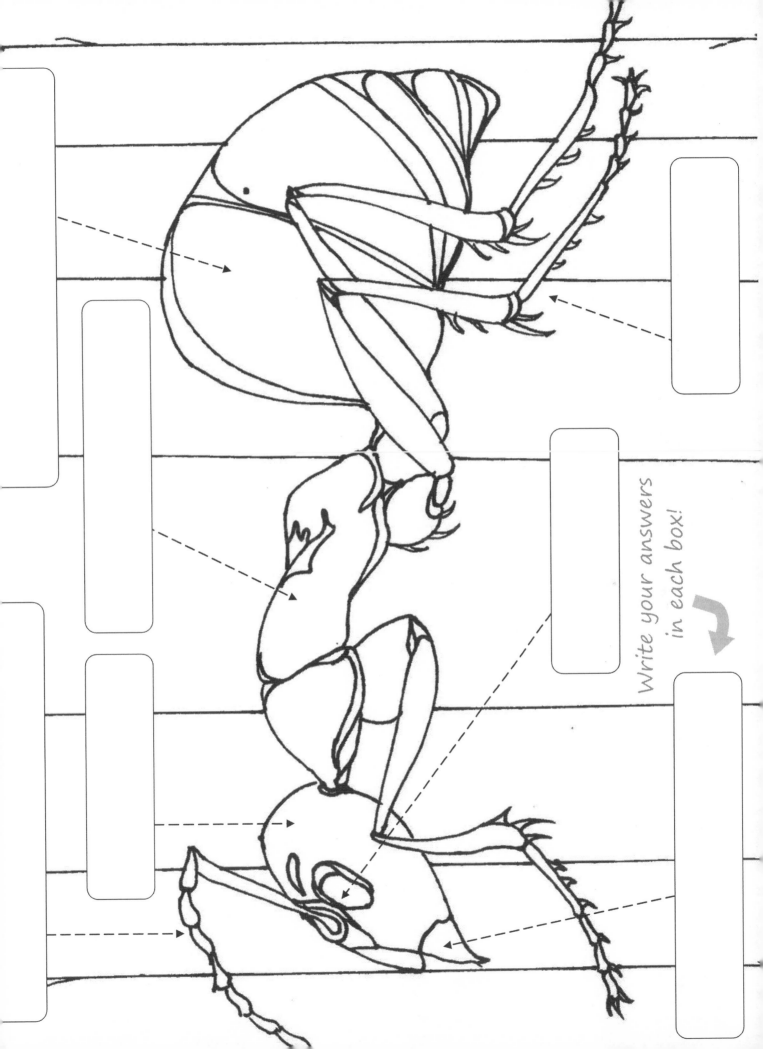

Write your answers in each box!

# Test tubes

are glass equipment found in laboratories with one end open and the other end closed. They can hold a small amount of substance and can be heated if needed.

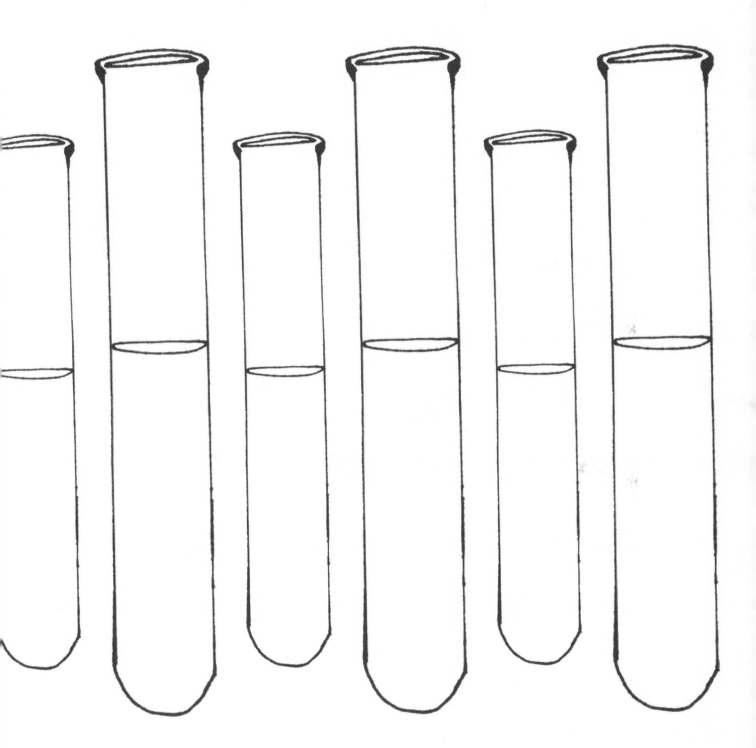

# The telescope

was first created in the 1600's to observe our nighttime sky. They can be made using either lenses or mirrors.

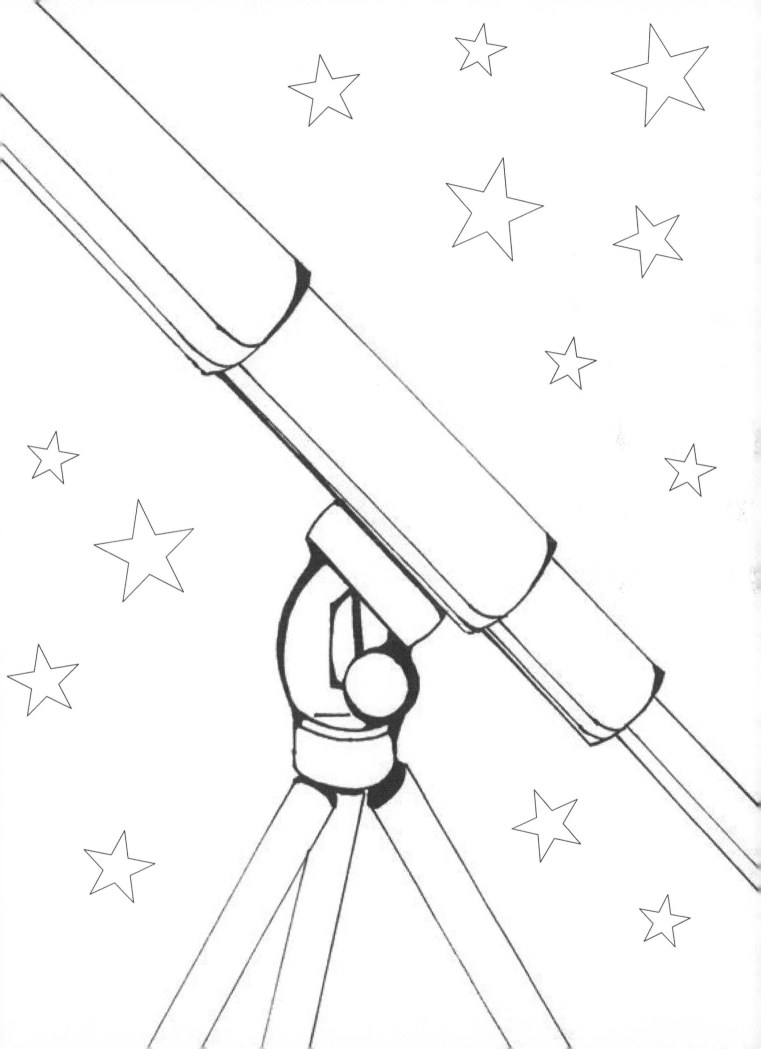

The cell is the building block of all life. Can you name its different parts as you color the next page?

Nucleus

Nucleolus

Endoplasmic reticulum

Golgi apparatus

Mitochondria

Vacuole

Cell membrane

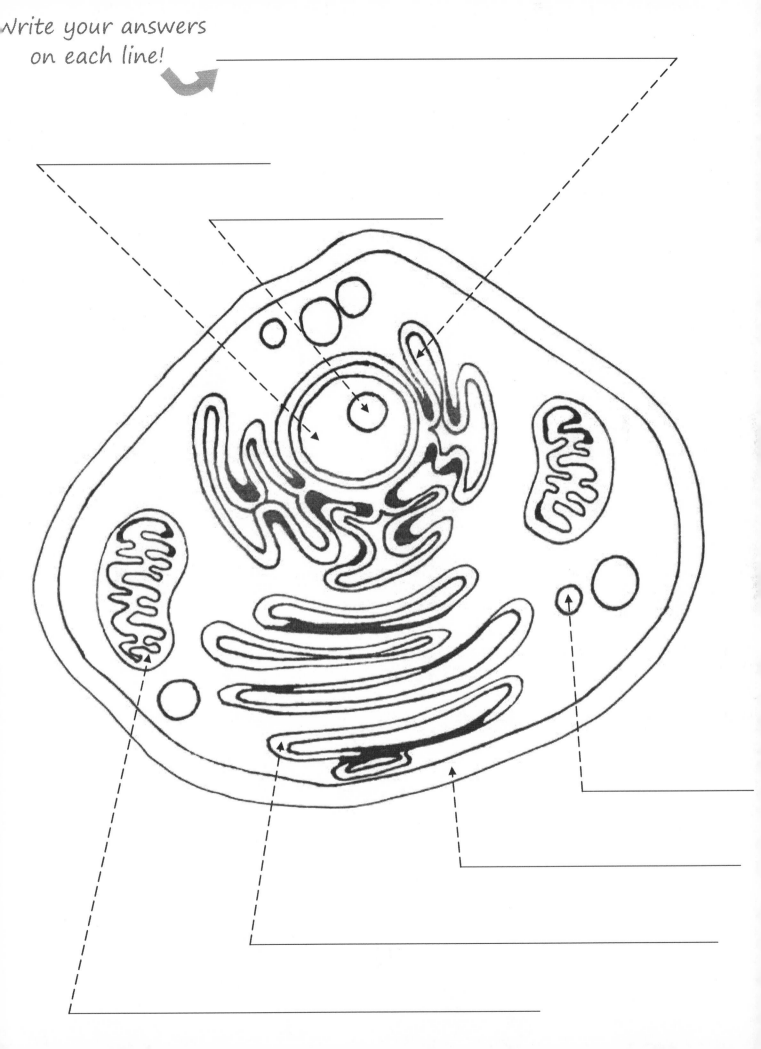

Write your answers on each line!

A circle is a basic geometric shape.
Can you define these terms as they apply to a circle?
Radius
Diameter
Circumference
Chord

# What is "$\pi$" ?

3.14159265358979793

2384626433832795

0288419716939937

5105820974944592

3078164062862089

9862803582534211

7067982148086513 2

8230664470938446 0

9550582231725359

1081284811117450 28

# An abacus

is an ancient tool used to perform mathematical calculations such as addition and subtraction. It is thought to have been used in various locations across the globe including Greece, China and Russia.

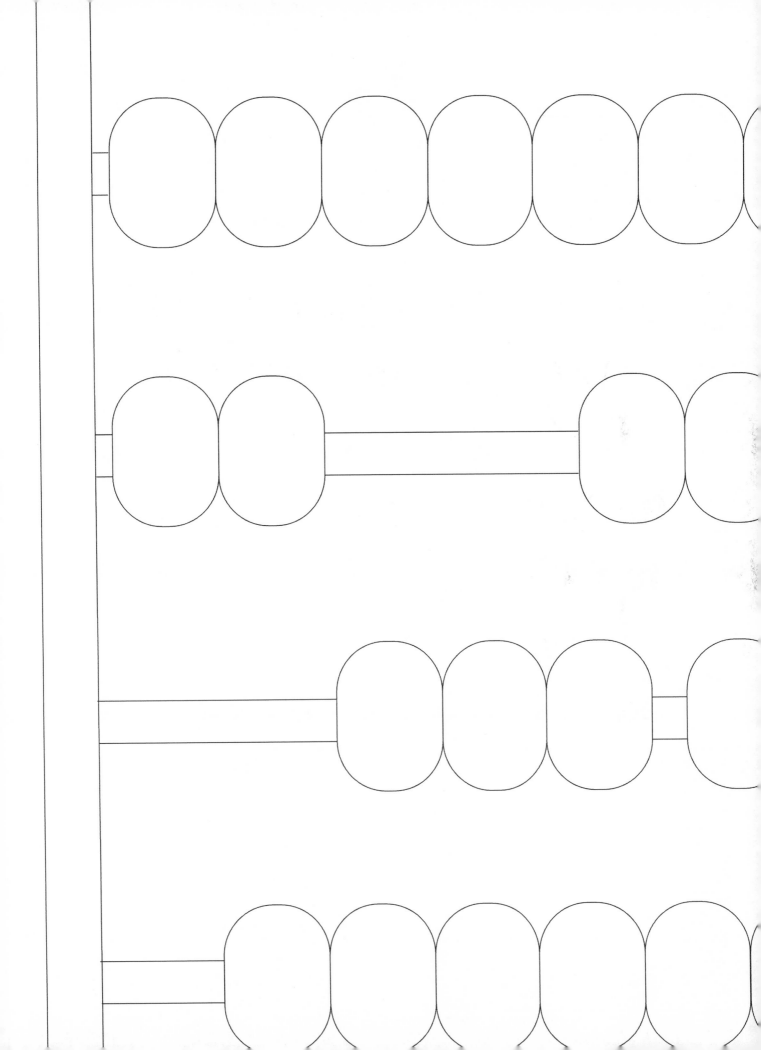

Everything around us is made of atoms. Can you name each part of the atom as you color the next page?

Proton

Neutron

Electron

Valence

Nucleus

## ABOUT THE AUTHOR

Thomai Dion is a pharmacist and mother to a very inquisitive, energetic and hands-on analytical thinker. She obtained her doctorate from the University of Rhode Island and believes it is never too early to start learning.

For more books that promote S.T.E.M. education, visit Thomai's Amazon author page:

## www.amazon.com/author/thomaidion

Continue on to the next page for this book's answer key!

Answer Key

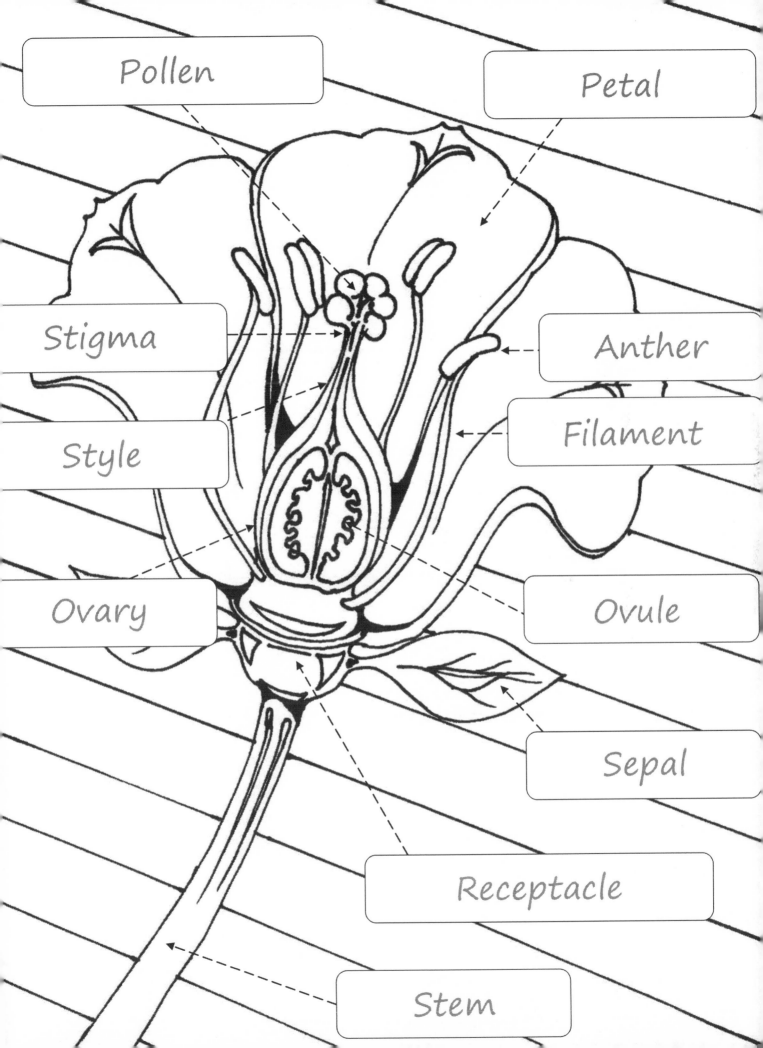

Pollen

Petal

Stigma

Anther

Style

Filament

Ovary

Ovule

Sepal

Receptacle

Stem

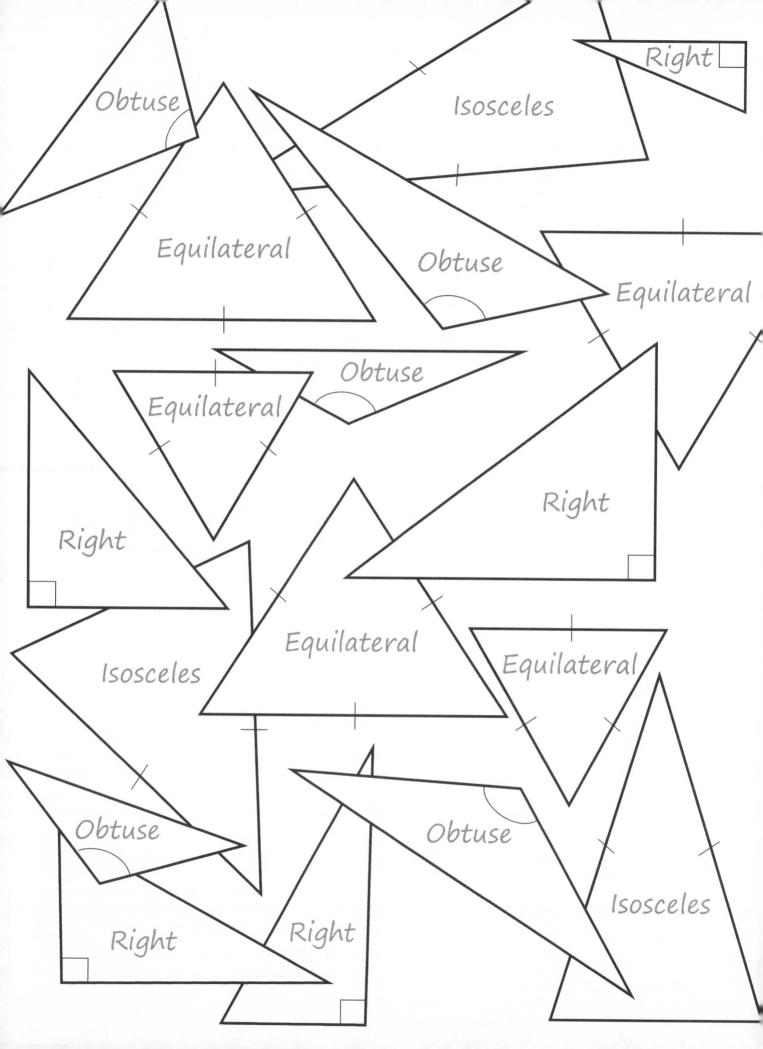

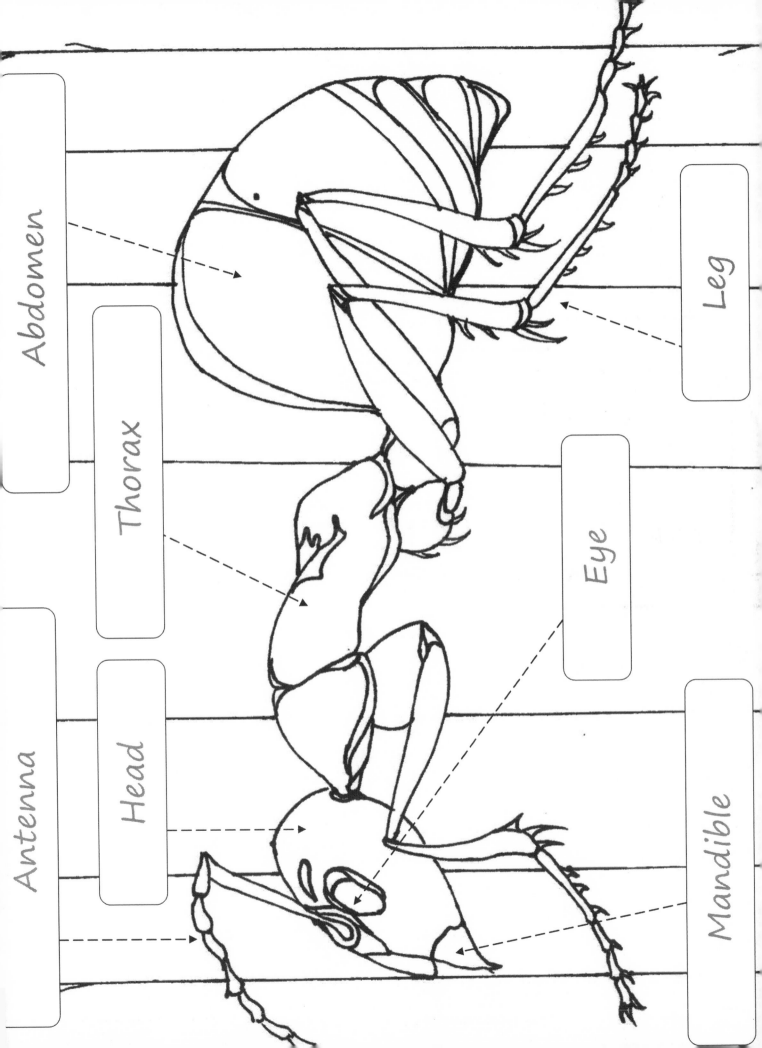

Abdomen

Thorax

Antenna

Head

Leg

Eye

Mandible

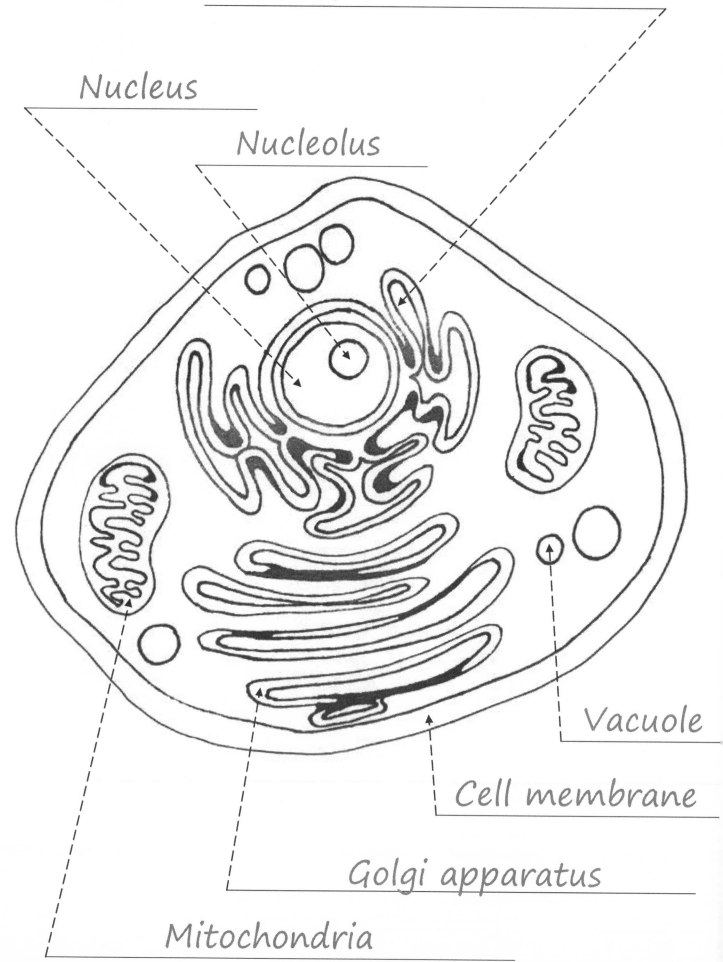

Endoplasmic reticulum

Nucleus

Nucleolus

Vacuole

Cell membrane

Golgi apparatus

Mitochondria

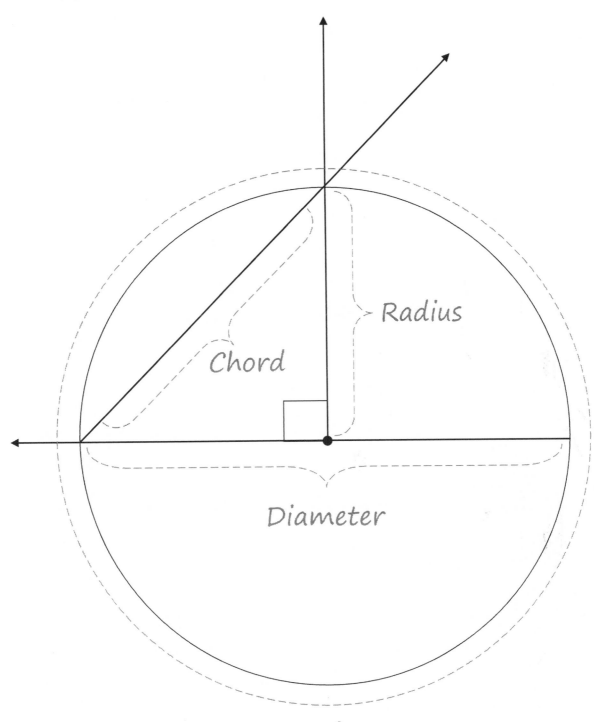

Chord

Radius

Diameter

Circumference

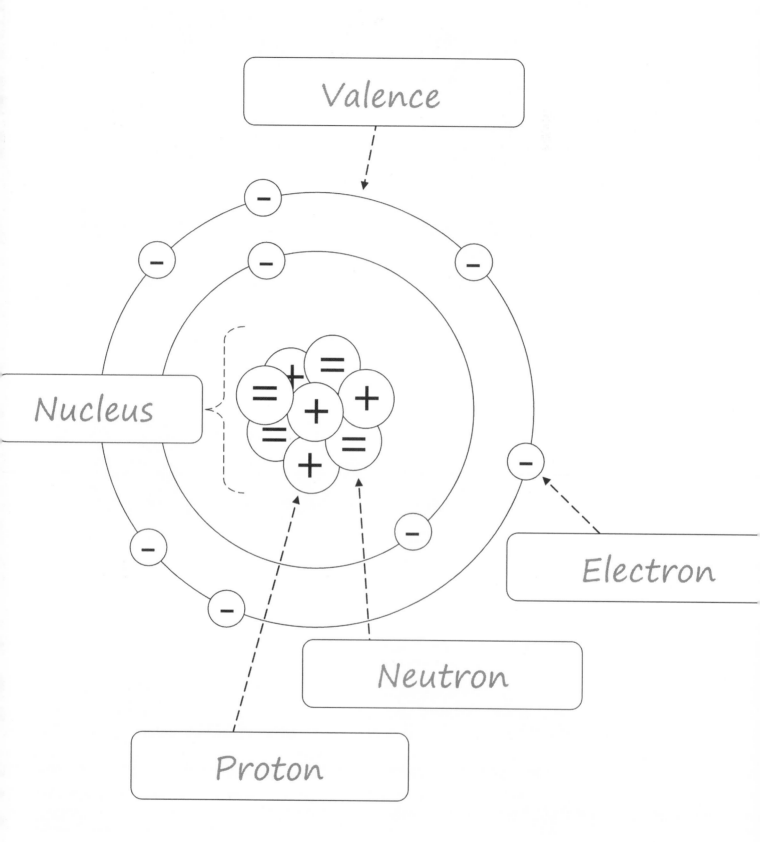

Valence

Nucleus

Electron

Neutron

Proton

# What does the acronym "S.T.E.M." stand for?

Science, technology, engineering and mathematics

# What is "π" ?

π ("pi") is defined as the ratio of the circumference (c) of a circle to its diameter (d).

When calculated (π = c/d), the result equals ~3.14, although the exact value of π is unknown.

Made in the USA
Middletown, DE
29 April 2018